RAND

Life Cycle Cost Assessments for Military Transatmospheric Vehicles

Mel Eisman, Daniel Gonzales

Prepared for the
United States Air Force

Project AIR FORCE

Preface

Transatmospheric vehicles (TAVs) are envisioned as new types of reusable launch vehicles that could insert themselves or payloads into low earth orbit or deliver payloads to distant targets within minutes. Such vehicles may be able to carry out military, civil, and commercial missions. In past decades, a number of military TAV concepts have been proposed, but a complete operational vehicle has never been built.

This report briefly reviews potential military missions that TAVs could perform and how a cost-effective DoD acquisition program could be initiated. We focus specifically on one promising military TAV design concept.

Research, development, test and evaluation (RDT&E) costs are estimated for the identified TAV design concept; they include production of both a demonstrator "X" and prototype "Y" vehicle during the development phase. A total life cycle cost (LCC) budget forecast is generated for one TAV and for a fleet of six operational military TAVs. The total number of TAV launches provided over the assumed service life of the vehicle fleet is compared to the number provided by an expendable launch vehicle for the same total projected budget. The costing methodology developed to arrive at these cost estimates is described along with suggested implementation strategies.

This work was done for the Future Role of the Air Force in Space project, one of several studies being carried out in the Force Modernization and Employment Program of Project AIR FORCE.

Project AIR FORCE

Project AIR FORCE, a division of RAND, is the Air Force federally funded research and development center (FFRDC) for studies and analyses. It provides the Air Force with independent analyses of policy alternatives affecting the development, employment, combat readiness, and support of current and future aerospace forces. Research is performed in three programs: Strategy and Doctrine, Force Modernization and Employment, and Resource Management and System Acquisition.

Contents

Preface iii

Figures vii

Tables ix

Summary. xi

Acknowledgments xix

Glossary xxi

1. REPORT OUTLINE 1

2. INTRODUCTION. 3
    Research Objectives 3
    Caveats 3
    Military TAVs—Why Now? 5

3. TAV AND RLV NEEDS AND CONCEPTS 7
    Military TAV Needs 7
        Military Missions 7
        Alternatives to Military TAVs 8
        Military Mission Effectiveness Trade Studies 8
        Potential Military Multimission Capability 10
    Commercial and Civil RLV Goals and Motivations 10
    Military TAV Technology Required 11
        Structures, Aerodynamics, and Engine Designs 12
        Propellant Mix and Fuel Handling 12
        Thermal Protection Systems (TPS) 12
        Flexibility and Operability 12
    Overview of RLV and TAV Concepts. 13
    Candidate Military TAV Design Concepts 13

4. MILITARY TAV COST ASSESSMENTS 17
    Ground Rules and Assumptions 17
    RDT&E Cost Assessments 19
        Cost Model Evaluations and Advantages of TRANSCOST 19
        Technology Readiness Levels and Risk Assessment Adjustments . . . 23
        Cost Estimating Cross Checks 24
        RDT&E Cost Results 25
    Overall LCC Assessments 26
        LCC Methodology 26
        LCC Affordability Assessment Results 26
        LCC Affordability Observations 28

5. OBSERVATIONS 30
    Summary of Cost Assessments 30
    Potential Implementation Strategies. 31

vi

Appendix
A. INITIAL MILITARY TAV AFFORDABILITY ASSESSMENT   33
B. RAND MILITARY TAV COST MODEL SAMPLE OUTPUT
   REPORT   40

Bibliography   45

Figures

S.1. Economic Advantages of Representative Military TAVs over
ELVs xvi
3.1. Representative Launch Vehicle Response Times 9
3.2. Representative TSTO Military TAV Cost Baseline Concepts 16
4.1. Sample TRANSCOST 6.0 RDT&E Engine CERs 21
4.2. Sample TRANSCOST 6.0 RDT&E Vehicle CERs (Excluding
Engines) 22
4.3. Technology Readiness Level (TRL) Definitions 23
4.4. Economic Advantages of Military TAVs over ELVs 29
B.1. Military TAV Cost Model Report, Input 41
B.2. Military TAV Cost Model Report, Output 43

Tables

S.1. Military TAV R&D Affordability Assessment (FY97 $M) xiii
S.2. Military TAV Life Cycle Cost Affordability Assessment xv
3.1. Recent RLV and TAV Concept Proposals 14
4.1. Military TAV Quantitative TRL Assessments 24
4.2. Military TAV RDT&E Cost Estimates (FY97 $M) 25
4.3. Military TAV Life-Cycle Cost Estimates (FY97 $M) 27
A.1. Initial Military TAV RDT&E Affordability Assessment 34
A.2. Initial Military TAV Life Cycle Cost Affordability Assessment 38

Summary

Background

The X-33 and X-34 technology demonstration programs currently under way by NASA may ultimately lead to a commercial reusable launch vehicle (RLV). The first flight test for the X-33 vehicle is planned for early 1999. With NASA fully committed to the development of this first-ever completely reusable vehicle, it is prudent for the Air Force to examine the potential benefits, requirements, and costs for a military transatmospheric vehicle (TAV). This vehicle would take advantage of technologies developed during RLV-related programs.

Although similar in concept, a military TAV may differ significantly in design and in capability from an RLV produced to satisfy commercial market needs.

Report Overview

The report covers three major topics relating to military TAVs. The first part describes significant differences between military TAVs and commercial RLVs in terms of mission objectives, maintainability, and response times—differences that could lead to different vehicle designs. Further mission assessment details are provided in a related RAND report.[1]

Several candidate military TAV design concepts were presented at an April 1995 RAND-sponsored workshop, including concepts with various launch and landing modes (i.e., vertical or horizontal takeoff or landing) and vehicle staging (i.e., two stage to orbit (TSTO) or single stage to orbit (SSTO) systems). Initial RAND assessments identified several horizontal takeoff, horizontal landing (HTHL) TSTO TAV concepts that at the time were observed to be better suited for military mission operations than the RLV designs submitted in the NASA X-33.SSTO Phase II competition.

The second part of the report briefly summarizes these concepts and covers some of the technical features of the TSTO TAV candidates. Further descriptions of the

---

[1]David Gonzales, Melvin Eisman et al., *Proceedings of the RAND Project AIR FORCE Workshop on Transatmospheric Vehicles*, RAND, MR-890-AF, 1997; see Section 2 of that report for further details.

overall RLV and TAV design options and issues are provided in Section 3 of MR-890-AF.

The third aspect—and primary focus of the report—describes the ground rules and assumptions, methodology, and rationale for estimating military TAV costs, with all the inherent features of reusability and aircraft-like operations that are separate and distinct from expendable launch vehicle (ELV) cost-estimating approaches. Preliminary RDT&E and overall life cycle cost (LCC) assessments on two candidate military TAV design concepts are provided. The report concludes with observations on these cost assessments from an affordability standpoint and thoughts on some alternative implementation strategies for entering this "new start" program into the Department of Defense (DoD) and Air Force acquisition process.

Military TAV Candidates

In August 1996, two of the TSTO vehicle concepts from the Northrop Grumman Corporation (NGC) were examined in detail as candidate military vehicle cost baselines after a RAND technical team review and assessment of the vehicles' aerodynamic characteristics and payload lift capabilities.

At that time, these TAV concepts appeared to be well suited for several military missions. Either vehicle could potentially deliver a 1000 to 6000 lb payload to various low earth orbits (LEO). One TSTO version could be launched from on top of a Boeing 747 and the other aerial-refueled from a KC-135 tanker, separated, and then launched. Both orbital vehicles resemble a scaled-down space shuttle and would have a significant cross-range capability.

After this research was conducted, a number of additional industry concepts presented to the Air Force Space Command (AFSPACECOM) led to the Military Spaceplane (MSP) Integrated Concept Team (ICT). The specific TSTO military concepts reported here are still under consideration along with the other TAV or MSP concepts. The detailed cost analysis of the NGC TAV concepts should be considered only as a proof of concept for the cost analysis methodology and is not an endorsement of the NGC TAV concepts over other TAV designs.

RDT&E Cost Assessment and Comparisons

A cost assessment for these concepts is described in detail in this report. The results represent an upper-bound estimate of the RDT&E budget required to complete a combined Demonstration/Validation (DEM/VAL) and Engineering,

Manufacturing and Development (EMD) phase for a military TAV program, including the delivery of a subscale X-vehicle and an operational prototype Y-vehicle.

The RDT&E costs include nonrecurring costs to perform design trades and those activities required to produce a technically feasible operational vehicle design. A production phase would follow with the manufacture and delivery of operational TAVs.

To gain an understanding of whether a military TAV would be affordable, we compared launch vehicle dry weights, estimated costs, and development complexity of an existing ELV, Pegasus, with two versions of the candidate military TAV. All the candidate vehicles can potentially handle at least some of the payload range requirements envisioned for a military TAV, although each of the vehicles may not satisfy all military needs.

The results of this cost analysis are summarized in Table S.1.[2] All costs are displayed in constant fiscal 1997 dollars. Using the NGC TSTO TAV as a representative basis for a military TAV, an upper bound of $760M is obtained for the aerial-refueled TAV variant and an upper bound of $680M is obtained for the air-launched TAV variant.

As displayed on Table S.1, the estimated cost of developing and testing a military TAV X and Y vehicle is significantly greater than the RDT&E estimated costs of

Table S.1

Military TAV R&D Affordability Assessment (FY97 $M)

Factor	Pegasus Expendable	TSTO Aerial-Refueled TAV	TSTO Air-Launched TAV
Total vehicle dry weight (lb)[a]	6,615	25,000	34,240
Payload weight to LEO (lb)	800	1,600 - 2,000[a]	3,000 - 6,000[a]
Total engine weight (lb)		4,535	3,000
Engine RDT&E cost ($M)[b,c]		129.0	87.0
Vehicle RDT&E cost ($M)[b,d]		630.0	590.0
Total RDT&E cost ($M)	149.0	759.0	677.0

[a]Weights assume an unmanned TAV version.

[b]Costs are based on a modified use of TRANSCOST 6.0, 1995 Edition (version 6.0) Cost Estimating Relationships (CERs).

[c]Rocket engine R&D costs are based on engine weight and an assumed 200 test firings required for flight certification and qualification.

[d]Vehicle R&D costs are based on vehicle dry weight (excluding engines), design maturity, and team experience.

---

[2]The cost model report details are provided in Appendix B.

$149M for Pegasus. However, the military TAV RDT&E estimates of $680M and $760M, even escalated to then-year dollars, are considerably less than the NASA X-33 contractor estimates of total development cost (which exceeds $1B) for a full-scale RLV. Thus, a relatively small military TAV that costs only about $700M to develop may be a reasonable investment toward achieving the responsiveness and flexibility needed for the military missions described in this report.

Overall Life Cycle Cost Assessments

We now compare the overall life cycle costs for a military TAV with a small ELV, again using Pegasus for "first order" comparison purposes.[3] The higher front-end RDT&E budget for a military TAV should result in lower total recurring costs relative to a comparable ELV such as Pegasus.

The recurring launch costs for an expendable Pegasus include the total production cost for each vehicle plus the fixed and variable launch operations cost. The total RDT&E estimated cost of $149M for Pegasus can be amortized over the expected number of operational ELVs procured.

The higher overall RDT&E budget for TAVs can be treated as nonrecurring and amortized over the total number of anticipated flights over the vehicle's service life. Because of reuse, the military TAV RDT&E budget can be amortized over a larger number of flights than Pegasus. The total recurring cost per military TAV consists of the total operations and support (O&S) cost estimate per launch. The total recurring cost per Pegasus is comprised of both this O&S cost per launch and the unit production cost of each expendable vehicle.

Two total budgets are generated for procuring one TAV and a fleet of six TAVs based on the lowest LCC of the two versions—the TSTO air-launched vehicle. To get a relative economic sense of the value of a military TAV total LCC budget, a comparable number of launches was computed for Pegasus using this lowest estimated TAV LCC budget as the expenditure ceiling. It is assumed that each TAV vehicle could be reused for an average of 100 flights, and a fleet of six reused for 600 flights over a ten-year service life.

---

[3]Pegasus is considered in this report to be an ELV "comparable" to the military TAV candidate concepts on the basis of similar payload lift capabilities. We recognize that the Pegasus configuration selected has a lower-end lift capability than the more recent experience of Pegasus XL. However, there was not enough technical information available at the time of this assessment to utilize the same methodology for the higher-end ELV. As a minimum, the Pegasus estimate can be viewed as a lower-bound cost comparison point for cross-check purposes.

The results are summarized on Table S.2.[4] The total LCC and budget for 100 launches of one operational air-launched military TAV is $1862M. This same total LCC budget would cover expenditures for only 64 Pegasus launches. The total LCC budget for a fleet of six air-launched reusable TAVs and 600 flights was computed to be $7787M. This same LCC budget covers the costs for about 505 aerial-refueled TAV launches and only about 286 Pegasus launches.

Table S.2

Military TAV Life Cycle Cost Affordability Assessment

Factor	Pegasus Expendable	TSTO Aerial-Refueled TAV	TSTO Air-Launched TAV
Total RDT&E cost[a]	$149.0	$759.0	$677.0
Avg. recurring unit production cost[a]	$17.0	$60.0	$55.0
Avg. refurbishment cost/launch[a]	N/A	$1.7	$1.5
Direct operations cost (DOC)/launch[a]	$4.7	$7.6	$6.1
Indirect operations cost/launch[a]	$3.1	$3.9	$3.7
Launch insurance/launch[a]	$1.9	(included in DOC)	(included in DOC)
Total recurring cost/launch	$26.7[b]	$13.6[c]	$11.3[c]
Total LCC TSTO air-launched budget for 1 vehicle	$1,862.0	← $1,862.0	← $1,862.0
Equivalent number of launches	64	79	100
Total LCC TSTO air-launched budget for 6 vehicles	$7,787.0	← $7,787.0	← $7,787.0
Equivalent number of launches	286	505	600

[a]All costs are computed based on modified use of CERs from the TRANSCOST cost model, 1995 Edition (version 6.0).

[b]The total recurring cost per launch for the expendable Pegasus vehicle includes the unit procurement cost of $17.0M and the recurring O&S cost per launch of $9.7M. This O&S cost consists of adding up the $4.7M, $3.1M and $1.9M listed in the table above.

[c]The total recurring cost per launch for two TAV versions includes only the total recurring O&S costs from the three cost elements listed within each column of the table above.

Clearly, in the long term, an air-launched military TAV is significantly more cost-effective than Pegasus. The cost advantage for this TAV becomes even more apparent when one compares the total LCC per launch to deliver a pound of payload to LEO for each of the launch systems.

The estimated costs per pound of payload to LEO are summarized in Figure S.1. Because Pegasus has a maximum payload size of only 800 lb, its payload cost per pound is about $34K/lb. The payload cost per pound for the aerial-refueled

---

[4]The cost model report details are provided in Appendix B.

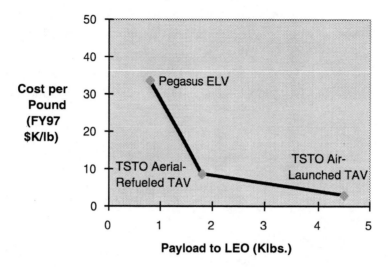

Figure S.1—Economic Advantages of Representative
Military TAVs over ELVs

TSTO TAV concept is estimated to be about $8.5K/lb. The payload cost per pound for the air-launched TSTO TAV concept is the lowest of the three systems at $2.9K/lb, which is more than an order of magnitude less than the Pegasus cost, and significantly less than the cost per pound of any launch vehicle available on the market today.

Observations

Military TAVs could provide competitive launch services for future commercial small satellite users. In the near term, many more small commercial remote sensing and communications satellites will be launched to serve emerging markets for mapping and geodesy information, imagery, and for cellular phone and Internet access services. Examples include Iridium, a constellation of 66 small satellites, and possibly Teledesic, a constellation of up to perhaps 900 small satellites. These emerging systems represent new business opportunities for small satellite launch service providers and may encourage significant new private investment in small TAVs.

Consequently, it may be possible for the Air Force and industry to share some percentage of the investment in RDT&E and vehicle production costs for small TAVs, if the program can be structured appropriately. It may even be possible for the government to share some operations and maintenance costs with industry. For example, if a TSTO TAV system were developed, the first-stage carrier aircraft could be shared or loaned to the Air Force on a basis similar to CRAF (Civil Reserve Aircraft Funding) during crises or war to meet surge requirements.

One example of an innovative financial structure for such a program is the NASA-industry cooperative agreement notice for the X-33 program, although a different financial model may be more appropriate in the small TAV case.

Perhaps a more appropriate cost-sharing arrangement is for DoD to underwrite early RDT&E investments and then share vehicle production start-up investment and capitalization costs. The first few TAVs produced would be delivered to the Air Force and their capabilities proven in military operations. Thus, the Air Force would make an up-front commitment to buy and own a small quantity of TAVs from the first-block build of vehicles. Later TAVs from the production line would go to the commercial launch service provider.

The guarantee of government revenue for procuring TAVs early in production provides a higher probability that corporate internal rate-of-return estimates and other financial goals can be met. By ensuring that corporate financial goals can be met, including the government-underwritten guarantee of sufficient cash flow, the TAV producer can keep the production line open and produce additional TAVs for commercial customers.

If the military needed to launch small satellites during crisis or war, the DoD could potentially take advantage of these commercial TAVs. Some commercial satellite launches may have to be rescheduled to allow for the timely launch of DoD space assets. However, because of the potential responsiveness of such TAVs, commercial launch schedules may not be seriously impacted, and regardless of what type of cost sharing arrangement is made, a "win–win" acquisition strategy for the Air Force and for industry might be implemented if an innovative acquisition approach can be adopted.

Acknowledgments

The authors thank Mr. Robert Haslett of the Northrop Grumman Corporation for sharing his company's work on several TAV design concepts and technical information that we used as a basis for a representative military TAV cost baseline; and Lt. Col. Jess Sponable, Maj. Mitchell Clapp (Retired), and Mr. Kenneth Hampsten of Phillips Laboratories for sharing their work on transatmospheric vehicles and related subjects with RAND.

Finally, we also thank Mr. Karl Hoffmayer and Mr. Charles Kelley for their expert technical reviews, Ms. Jan Harris and Ms. Elaine Wagner for their assistance in the preparation of various stages of this report, and Ms. Jeanne Heller for the final editing of this manuscript.

Glossary

AFPL	Air Force Phillips Laboratory
AFMC	Air Force Material Command
AF/SMC	Air Force Space and Missiles Center
AFSPACECOM	Air Force Space Command
AIAA	American Institute of Astronautics and Aeronautics
AOR	area of operational regard
ASATs	Anti-Satellite Weapons
CRAF	Civil Reserve Air Fleet
CERs	cost estimating relationships
CONOPS	Concept of Operations
DEM/VAL	Demonstration/Validation
DoD	Department of Defense
EELV	Evolving Expendable Launch Vehicle
ELV	Expendable Launch Vehicle
EMD	Engineering Manufacturing and Development
FFP	firm fixed price
FFRDC	Federally Funded Research and Development Center
GEO	Geosynchronous Earth Orbit
GPS	Global Positioning System
GR&As	ground rules and assumptions
GTO	Geosynchronous Transfer Orbit
HLV	Heavy Launch Vehicle
HTHL	horizontal takeoff, horizontal landing
ICBM	Intercontinental Ballistic Missile
IRR	internal rate of return
LCC	life cycle cost
LEO	Low earth orbit
LOX	liquid oxygen

LOX/LH2	liquid oxygen/liquid hydrogen
LOX/RP	liquid oxygen/rocket propellant
LVCMs	Launch Vehicle Cost Models
MLV	Medium Launch Vehicle
NASA	National Aeronautics and Space Administration
NASP	National Aerospace Plane
NGC	Northrop Grumman Corporation
OSC	Orbital Sciences Corporation
O&S	operations and support
POST	Program to Optimize Simulated Trajectories
PRC	Planning and Research Corporation
RASV	Reusable Aerodynamic Space Vehicle
RDT&E	Research, Development, Testing, and Evaluation
ROM	rough order of magnitude
RLV	Reusable Launch Vehicle
RV	Reentry Vehicle
SSTO	Single Stage to Orbit
TAV	Transatmospheric Vehicle
TPMs	technical performance measures
TPS	thermal protection systems
TRANSCOST	Model for Space Transportation Cost, Estimation, and Economic Optimization
TRD	Technical Requirements Document
TRLs	Technology Readiness Levels
TSTO	Two Stage to Orbit
VTVL	Vertical Takeoff, Vertical Landing
WMD	Weapons of Mass Destruction

1. Report Outline

The following section headings indicate topics covered in this report:

- Introduction
- Transatmospheric Vehicle (TAV) and Reusable Launch Vehicle (RLV) Needs and Concepts
- Military TAV Cost Assessments
- Observations.

The introduction (Section 2) describes (1) the specific research objectives within the Project AIR FORCE Future Roles of the Air Force in Space project; (2) major caveats applicable to this research; and (3) why military TAVs should be considered today.

Section 3 discusses (1) Air Force involvement in RLVs; (2) potential military missions a TAV could perform; and (3) how a cost-effective Department of Defense (DoD) acquisition program could be initiated. We contrast and compare military and civil/commercial needs and associated designs for launch vehicles. RLV concepts are listed and the features needed for possible military TAV candidates are described. Two specific candidates are selected and described as representative conceptual designs for a proof-of-concept of the cost methodology.

Costs for the two military TAV conceptual designs are assessed in Section 4, and the applicable cost ground rules and assumptions (GR&As) and the cost methodology are described. The cost assessment results include a range estimate of RDT&E costs and overall life cycle costs (LCC) for both representative concepts. For each military TAV concept, a total LCC budget is computed over an assumed service life. As one measure of effectiveness, the number of equivalent launches of the representative military TAV is compared with an expendable launch vehicle (ELV) with an equivalent budget and similar payload lift capability. These preliminary results provide an assessment of the economic advantages of military TAVs over ELVs.

Section 5 provides some observations of (1) the overall cost assessment results including the LCC budget affordability range estimate; and (2) potential implementation strategies for a program within the DoD and Air Force

2

acquisition process. These strategies are described in the context of how best to proceed given the timelines and budgets of the current NASA X-33 and X-34 RLV programs and the DoD Evolved Expendable Launch Vehicle (EELV) program.

2. Introduction

Research Objectives

The research objectives of one aspect of the Future Roles of the Air Force in Space project include

- a review of potential military missions that TAVs could perform;

- a list of enabling technologies to take advantage of;

- differences between civil/commercial RLVs and military TAVs; and

- rough order of magnitude (ROM) RDT&E and LCC assessment implementation strategies.

The primary focus of this report is to document the preliminary cost assessments and implementation strategies.

Caveats

The military TAV baseline design concepts described in this report should be considered as proof-of-concept benchmarks for the cost analysis methodology. The concepts should not be considered as an endorsement over other TAV designs. They are based on a selection of contractor concepts available through FY96 that can lift "nominal" payloads into low earth orbit (LEO) or suborbitally to specific theaters of operation.[1]

At this time, the initial cost estimates provided here are limited to the primary vehicle itself based on our preliminary assessment of military mission needs, candidate conceptual platform designs to meet the projected suborbital and orbital flight profiles, and "nominal" payload lift capabilities. In addition, all candidate concepts assessed are configured as **uninhabited** with no life support

---

[1]The "nominal" payload lift capability for the military missions described in this report are estimated to be between 1000 and 6000 lb. Indications from the Air Force Space Command Integrated Concept Team (AFSPACECOM ICT) could push some global reach missions beyond the 6000 lb upper limit, which could force a different set of design solutions. However, even this upper-bound payload weight is still less than the civil and commercial RLV projected lift demands of between 25,000 and 40,000 lb.

system and onboard displays required.[2] Given the fidelity of platform design details available and the confidence level of the estimating methodology, the point estimates are accurate within ± 20 percent at the 95 percent confidence level.[3]

We recognize there are other subsystem cost elements that need to be included. However, since specific mission-related configuration design details, force mix quantities, and acquisition priority levels are highly uncertain, the cost assessments provided here **exclude** estimates for mission payload development and production costs, assembly, integration and test (AI&T) costs for vehicle payload, and related ground mission operations and support (O&S) equipment costs. Attempts to add these estimates to the platform vehicle–level estimates will only further increase the uncertainty and bandwidth of the range estimates. Nonetheless, the economic advantages of the reusable vehicles over mission-comparable ELVs can still be evaluated on a relative platform-to-platform cost basis.

Finally, for purposes of these preliminary cost assessments, a traditional acquisition program of a combined risk reduction or Demonstration/Validation (DEM/VAL) and Engineering Manufacture and Development (EMD) phase is assumed prior to go-ahead into production. This acquisition approach appears to be more representative of an upper-bound budgetary baseline. However, we recognize that this acquisition approach is only one of several possible alternatives, and it may not necessarily reflect the planned acquisition strategy that will be implemented within DoD and the Air Force for a military TAV program.[4]

Given these caveats, we recognize the need to update these cost assessments using the cost methodology to more appropriately reflect a total budgetary projection based on the latest mission-specific information and acquisition guidance available from the Air Force Military Space Plane (MSP) ICT and other related activities.

---

[2]We recognize that a piloted vehicle is still a viable option within the overall military TAV solution space. However, for simplicity we chose to select only uninhabited vehicles to minimize the potential operational complexity of the vehicle and keep the concept technically similar to ELVs for cost comparison purposes.

[3]The point estimate accuracy is based on a 95 percent or 2-σ confidence level assuming a normal distribution.

[4]As an example, depending on the assumed extent of mature technology and total budget available to meet a projected military TAV operational time frame, future cost assessments could consider a more streamlined acquisition reform approach.

Military TAVs—Why Now?

There are two separate but related questions on why military TAVs should be considered now.

1. Is there military utility in having TAVs as part of the future force mix? Will this result in a more cost-effective approach for future applicable military missions?

The first part of this question can be answered in the affirmative. There are at least three potential military missions TAVs may be well suited for:

- **space forces support** (space lift of communications, navigation, and reconnaissance and surveillance satellites);

- **space control** (ensuring friendly use of space while denying its use to the enemy); and

- **space force application** (attacks against weapons operating in or through space).

The three military missions are described further in Section 3. A more detailed treatment is provided in MR-890-AF.[5]

Since these three broad mission areas were identified in August 1996, further missions have been identified by the AFSPACECOM ICT that appear to be subsets of the three major missions listed above.[6]

The answer to the second part of the question depends, in part, on the priority of mission needs and on the overall implementation approach that will be taken within the projected program budget and schedule.

2. Is the technical maturity of the industry's military TAV concepts far enough along to provide sufficient readiness to respond to the future performance, operational, and supportability needs?

---

[5]Daniel Gonzales, Melvin Eisman, et al., *Proceedings of the RAND Project AIR FORCE Workshop on Transatmospheric Vehicles*, RAND, MR-890-AF, 1997; see Section 2 of that document for further details.

[6]An AFSPACECOM Statements of Need document identified six potential mission areas where a military space plane or TAV may be required. As part of space launch support, a TAV could possibly function as a low-cost space hangar for facilitating in-space operations or as a platform for on-orbit repair of satellites. Space control could include a TAV that provides global reach for precision viewing of targets. Space force applications could include the prompt neutralization of adversarial satellites without creating space debris. In addition, space force could apply to precision strike attacks through use of nonnuclear multiple independently targeted reentry vehicles (MIRVs) released from space and the upper atmosphere.

Military TAVs can take advantage of the current NASA X-34 and X-33 Phase II technology demonstration programs for development of civil/commercial RLVs. The X-33 demonstrator's first flight test is scheduled for early 1999. Military TAV design concepts may also benefit from enabling technology developed during previous RLV-related programs such as the

- Reusable Aerodynamic Space Vehicle (RASV),
- National Aerospace Plane (NASP), and
- DC-X and DC-XA (Delta Clipper).

Applicable technology transfers from the civil and commercial RLV area are briefly described in Section 3. A more detailed treatment may be found in MR-890-AF. For the representative military TAV concepts selected, technical readiness levels and associated risk assessments at the subsystem and below levels are described in Section 4 as part of the approach taken to size the nonrecurring RDT&E effort and costs required.

In summary, the responses to the above questions indicate that it appears very timely for the Air Force to examine **now** the potential benefits, requirements, and costs of military TAVs.

3. TAV and RLV Needs and Concepts

Military TAV Needs

Since FY 1994, the Air Force Phillips Laboratory (AFPL) Advanced Spacelift Technology Program has assisted NASA in certain aspects of the X-33 and X-34 programs. In February 1995, the Advanced Spacelift Technology Program drafted a document that provides an initial set of technical requirements as guidance for the design of military TAVs.[1] In addition, AFPL has initiated interchanges with several Air Force users including the AFSPACECOM and the Air Combat Command to solidify mission requirements for a military TAV. A TAV Concept of Operations (CONOPS) document has recently been approved by AFSPACECOM.

Military Missions

After extensive discussions at a RAND-hosted TAV Workshop in April 1995 and after reviewing the AFPL-generated TRD, we believe there are significant differences between the mission and operational needs implied or assumed for a military TAV and those for the NASA RLV programs.[2]

A military TAV should be capable of handling space launch support, space control, and force application missions. In **space forces support**, a reusable military TAV could quickly deploy surveillance or communications satellites on short notice to fill coverage gaps in a specific area of operational regard (AOR). In addition, if a critical military satellite malfunctioned during a crisis situation, it could be replaced by another satellite deployed by a military TAV. The military TAV could also recover the damaged satellite and return it to earth for possible repair, if the satellite was small enough to fit within the military TAV payload envelope.

A military TAV could have significant military utility in **space control**, especially with the increasing capabilities of commercial satellites. Adversaries may be able to turn new commercial space assets to their advantage on the battlefield. A

---

[1] *Technical Requirements Document* [TRD] *for a Military TAV*, prepared by the AFPL for the Air Force Space and Missiles Center (AF/SMC), February 1995.

[2] See MR-890-AF, Section 2.

military TAV could deploy space control payloads during wartime to deny enemy access to satellite systems that provide coverage of the theater of operations.

The third mission area, **space force application**, is the attack of time-sensitive targets (e.g., invading armored columns or mobile ballistic missile launchers). A highly responsive military TAV may be able to deter or counter the use of such weapons or forces.

Alternatives to Military TAVs

Military TAVs are not the only alternatives that can satisfy these missions. Instead of launching satellites on demand with military TAVs for space support, satellites can be stored on orbit using ELVs. These on-orbit satellites can be activated and deployed to an appropriate orbital slot in crisis conditions.

Space control missions can be performed using ground-based antisatellite weapons (ASATs) and various countermeasures. Similarly, weapons can be delivered by bombers, cruise missiles, and conventionally armed ICBMs.

In the space force application mission area, the United States currently has limited space-related conventional capabilities. However, with the drawdown of U.S. forces forward-deployed overseas and the increasingly threatening long-range strike capabilities of potential adversaries, there has been increased emphasis on acquiring improved global strike capabilities. Space-delivered weapons have unique advantages over conventionally delivered weapons because they apply very high kinetic energy and fast closing velocities for improved survivability, responsiveness, and target lethality. However, these space-delivered hypersonic weapons require increased RDT&E investments to improve the guidance, maneuverability, and release of submunitions and to reduce recurring production costs by improving the producibility aspects of the unit.

Therefore, before drawing any conclusions, a thorough assessment of military TAVs versus other platforms capable of performing the three specific missions should be completed to determine the most cost-effective and responsive option for a broad range of weapons and payload sizes.

Military Mission Effectiveness Trade Studies

To carry out military space support missions, TAVs would have to be more responsive, flexible, and cost-effective than existing ELVs and provide significant operational advantages. As we shall show in the overall LCC assessment, military TAVs, because they would be reusable, would likely be more cost-effective than existing ELVs.

Responsiveness is a key operational characteristic of a military TAV. The TRD cited above states that a reusable vehicle should be ready for launch within seven calendar days under "normal conditions," and should be capable under "emergency or surge conditions" of doubling the flight rate and be ready for launch within hours. As seen in Figure 3.1 below, this type of responsiveness is closer to the capability of Pegasus than to that of traditional heavy and medium lift ELVs.

A RAND analysis of the space force applications (global reach) mission compared response times, lethality, mission complexity, asset resource allocations, deployment operations, and personnel. Three strike package options using military TAVs, F-117s, or B-2s as the strike platform were analyzed against the above criteria.

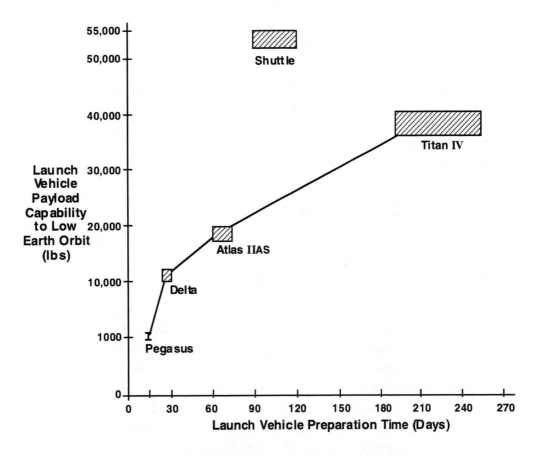

Figure 3.1 — Representative Launch Vehicle Response Times

Potential Military Multimission Capability

Military TAV concepts that are responsive, reusable, and can operate under unscheduled launches will most likely be beneficial and critical for the majority of missions listed above. Besides responsiveness for specific missions, the military TAV should also be flexible, since single-mission concepts may not be cost-effective, especially over the long run, in terms of total LCC.

The cost assessments presented later in this report assume the development of design concepts for a "family of military TAVs" that are rapidly reconfigurable with a containerized type of payload bay. Payloads could be reconfigured and changed quickly based on military mission requirements. With remove-and-replace modules for mission-unique payloads, a common platform with standard payload interfaces is designed to respond to the most stressing performance conditions across the missions.

This concept, if technically feasible, would reduce the number of "new start" programs required and, depending upon the number of years of front-end development activity required, could be more affordable across the projected budget than the sum of the total budgets required for single-mission approaches.

Commercial and Civil RLV Goals and Motivations

Civil/commercial RLVs are designed

- to reduce costs to compete effectively in the world market;
- to perform to scheduled launch manifests;
- to have a payload sized to the market; and
- to have preplanned flight profiles without major anticipated changes.

Each of these points will be expanded on below.

The dramatic decline in U.S. market share in the international commercial launch services market has caused growing concerns over the competitiveness and costs of the current fleet of U.S. ELVs. The goal of the NASA X-33 program is to develop a commercially viable RLV with sufficiently low operations costs to enable U.S. industry to "leap-frog" the foreign competition. The program is structured as a cooperative agreement in which NASA and U.S. industry jointly define the requirements and share development costs. Currently, Lockheed-Martin "Skunk Works" is in Phase II of this program.

During the RAND-hosted TAV Workshop, there was agreement among the Phase I X-33 contractors that certain DoD missions of greater than 45,000 lb going into geosynchronous earth orbit (GEO) that fall within the Titan-IV heavy launch vehicle (HLV) class are outside the practical design limits for a marketable RLV SSTO concept. The majority of the Delta and Atlas medium-class payloads of between 20,000 to 45,000 lb would be the market for the RLVs. NASA may choose a commercial RLV for space station replenishment and an RLV eventually could serve as the basis for a shuttle vehicle.

There is a definite difference when one compares commercial RLV mission needs with potential military TAV needs. An unpredictable launch manifest or launch on alert will be the desired TAV mode of operations compared with scheduled manifests for commercial customers. Many potential military mission needs could be satisfied by smaller payloads of 5000 lb or less.

Maneuverability is not as critical for commercial needs, and therefore the RLV can be designed with a low lift-to-drag ratio for a minimal amount of required cross-range movement. However, the flight profiles of some military missions would require the TAVs to have significant cross-range maneuverability. Military missions may also require the TAV to have the operational flexibility to land at multiple recovery sites.

Consequently, the combination of a rapid launch-on-alert capability, unpredictable launch schedule, fast turnaround time, and rapid reconfigurability to handle a variety of missions results in a set of requirements that are uniquely military. Designing a military vehicle to operate from a dispersed and survivable launch infrastructure would be more difficult than designing a civil and commercial RLV that has a highly structured and more predictable launch schedule and could operate out of one or possibly two launch sites.

In summary, even though there is some potential enabling technology available from civil/commercial RLV designs, there are unique demands placed on military TAV designs that may drive program costs and overall affordability. The next subsection and the cost assessments that follow reflect commercial/military differences.[3]

Military TAV Technology Required

Some of the potential RDT&E tasks for developing military TAVs are described below.

---

[3]See MR-890-AF; details on military TAV technology challenges are described in Section 4.

Structures, Aerodynamics, and Engine Designs

From an aerodynamic maneuverability standpoint, the lifting-body design of the X-33 may not have the cross-range capability required to perform military missions. In addition, varied mission needs of military TAVs will require a flexible yet integrated structural and aerodynamically sound design approach in which strict weight margins and mass fractions will have to be maintained. As part of system engineering, coordinated subsystem product teams should support early trade studies to achieve the highest feasible engine thrust-to-weight ratios for military TAV missions.

Propellant Mix and Fuel Handling

Development of a military TAV will require a thorough investigation of noncryogenic high-density propellants such as the combination of methane and liquid oxygen, which is a potentially inexpensive option. The performance of various noncryogenic fuel alternatives would have to be traded off along with their effects on RDT&E and launch operations costs.

Thermal Protection Systems (TPS)

In the area of TPS, vehicle development tradeoffs between manufacturing costs and launch operations costs may be required. Metallics and composites will have to be evaluated in terms of durability and reliability and for vehicle cross-range maneuverability and turnaround times. As part of the airframe and structural design, the program will have to analyze and test the cost-effectiveness of different robust advanced TPS that can withstand worst-case temperatures during various military flight ascent and reentry profiles.

Flexibility and Operability

The military TAV program will also have to assess the RDT&E and production cost impacts of designing in the flexibility of the vehicle to operate independent of traditional range safety constraints. Potential increased front-end RDT&E costs of designing in this capability will have to be compared and traded off with possible downstream launch operations and infrastructure cost savings resulting from reduced launch delays and increased availability of the vehicle to meet mission needs.

Finally, in cases requiring missions to abort, the military TAV program will have to assess the technical impact and increased RDT&E and production cost of designing in the capability of the vehicle to safely land at remote sites to allow maximum operational flexibility and reusability. Some of the flight test experiences gained from the DC-X , DC-XA, and upcoming X-33 and X-34 programs will be of benefit only where the mission profiles and trajectories are similar to the military TAV missions. Military TAVs will have more landing site abort options than commercial RLVs and may require different mission abort demonstrator tests.

Overview of RLV and TAV Concepts

Table 3.1 summarizes the characteristics of some of the RLV and TAV concepts that have been proposed over the last several years. Much of this information is from the RAND-hosted TAV Workshop. Three SSTO concepts for X-33 were proposed during Phase I, all with either different takeoff or landing approaches or different aerodynamic structural designs. For the two stage to orbit (TSTO) vehicle concepts, the majority had as a first stage a conventional carrier aircraft (e.g., KC-135, Boeing 747, etc.) with the RLV either air-dropped, air-launched, or aerial-refueled prior to reaching suborbital or orbital trajectories.

Candidate Military TAV Design Concepts

Initial observations from the conceptual information gathered through August 1996 indicated some observed general differences in vehicles depending on the launch and landing modes:[4]

- Single stage to orbit (SSTO) TAVs for military use provide single vehicle simplicity with less robustness and a more risky development path than TSTO platforms.

- TSTO TAVs improve payload performance and reduce overall vehicle dry weight over SSTO comparable platforms.

- TSTO vehicles with vertical takeoff provide more-robust performance than horizontal versions, but with additional complexity.

- Horizontal takeoff TSTOs show some potential to reduce complexity over vertical versions and provide aircraft-like operations.

---

[4]These are first-look observations based on limited data that require updating. These observations may be too general and should be reevaluated based on the unique conceptual designs that a number of contractors have presented to the AFSPACECOM MSP ICT over the last year.

14

Table 3.1

Recent RLV and TAV Concept Proposals

Vehicle	Contractor/Lab	Staging	Payload	Propulsion	Comments	
X-33	Lockheed Martin	SSTO	Heavy	LOX-LH2	Lifting body, VTHL, aerospike engine	■
X-33	Rockwell	SSTO	Heavy	LOX-LH2	VTHL	■
X-33	McDonnell	SSTO	Heavy	LOX-LH2	VTVL	■
X-34	OSC	Air-drop	Small	LOX-storable	HTHL, L-1011	■
REFLY	Rockwell	Air-drop Pegasus	Very small	Noncryogenic	L-1011, B-52, reusable upper stage	■
NG TAV	Northrop Grumman	Air-launched	Small	LOX-LH2	Boeing 747	
Black Horse	Phillips Lab	Aerial-refueled	Small	H₂0₂-Kerosene	KC-135Q tanker	☑
Neptune	Phillips Lab	Air-drop	Small	LOX-LH2	B-1B	
TAV	AMC HQ (Snead)	Air-launched	Medium	LOX-LH2	Boeing 777	☒

■ Under development (NASA) ■ Design proposed ■ Concept proposed

☑ Concept performance verified ☒ Concept performance problem identified

SOURCE: MR-890-AF, Table 3.2.

Consequently, we selected several technically feasible horizontal takeoff, horizontal landing (HTHL) TSTO concepts with robust designs to meet military mission needs, that are at the lower end of complexity and RDT&E costs, and that could perform in close to an aircraft-like environment to minimize operating and support costs and overall LCC.

Initially, we selected the AFPL Black Horse TSTO concept proposed in a Air University report as an initial representative cost baseline concept for a military TAV.[5] This initial cost assessment using the characteristics and parameters from the Black Horse concept is provided in Appendix A.

In addition, RAND performed a performance analysis on the NGC TAV and was able to verify the contractor's claimed payload delivery capability using a NASA trajectory analysis program, POST (Program to Optimize Simulated Trajectories). Analysis of the NGC TAV[6] confirmed the technical feasibility of the NGC design to deliver payloads of between 1000 to 6000 lb to various low earth orbits.

Both the NGC-proposed air-launched TSTO version that could be launched from the top of a Boeing 747 and the version aerial-refueled from a KC-135 tanker are considered worthy cost baseline design candidates for this study. Both have sufficient lift to reach orbit with their specified payload capabilities. These promising vehicle concepts appear well suited for several military missions.

Figure 3.2 is a schematic of the two proposed NGC RLV concepts on which the cost assessments are based. Both concepts are subscale versions of the space shuttle; the aerial-refueled version is slightly larger with more wingspan, length, and an extra LOX/RP (liquid oxygen/rocket propellant) engine added for additional thrust. Both orbital vehicle concepts use rocket propulsion and not air-breathing engines. The X-34 configuration would be air-launched from a Boeing 747 and the aerial-refueled version would be refueled in flight using a tanker aircraft.

As mentioned earlier in the report, even though NGC proposed these concepts as piloted, we compute payload lift capabilities and generate cost assessments assuming both concepts are uninhabited. The air-launched version with only two engines has a better thrust-to-weight ratio and can lift more equivalent

---

[5]Some of the Black Horse technical description was obtained from Maj. Chris Daehnick (USAF), "*SPACE LIFT Suborbital, Earth to Orbit, and on Orbit* (A SPACECAST 2020 White Paper)," Air University, published in *Air Power Journal*, Summer 1995.

[6]Technical data on the NGC TAV concepts were based on discussions and a briefing by Robert Haslett of the Northrop Grumman Corporation on "TAV Concepts Briefing," presented at the TAV Workshop, April 1995.

16

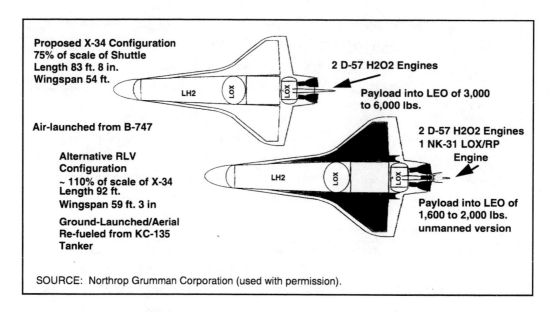

Proposed X-34 Configuration 75% of scale of Shuttle Length 83 ft. 8 in. Wingspan 54 ft.

2 D-57 H2O2 Engines

Payload into LEO of 3,000 to 6,000 lbs.

Air-launched from B-747

Alternative RLV Configuration ~ 110% of scale of X-34 Length 92 ft. Wingspan 59 ft. 3 in

Ground-Launched/Aerial Re-fueled from KC-135 Tanker

2 D-57 H2O2 Engines 1 NK-31 LOX/RP Engine

Payload into LEO of 1,600 to 2,000 lbs. unmanned version

SOURCE: Northrop Grumman Corporation (used with permission).

Figure 3.2— Representative TSTO Military TAV Cost Baseline Concepts

payload into LEO than the aerial-refueled concept. The aerial-refueled version will have a lower takeoff weight and will refuel the oxidizer from a tanker carrier before proceeding to orbit.

NGC's air-launched concept would have a cross-range of 1000 nmi and is designed for a minimum lifetime requirement of 100 flights. The mission profile allows for a 24-hour maximum time in a single orbit. The D-57 is a Russian engine that is being licensed for production through Aerojet. This engine is rated at 452 seconds of specific impulse and uses a combination of LOX and LH2 propellants. The engines are fully throttleable. More than 105 engines have been built and significant test firings have been conducted by AFPL.

4. Military TAV Cost Assessments

Ground Rules and Assumptions

The following ground rules and assumptions (GR&As) apply for both the RDT&E and overall LCC assessments. Note that LCC estimates include RDT&E costs.

1. Cost assessments identify upper-bound budgetary estimates of the total RDT&E and LCC costs required to complete a traditionally structured acquisition of a combined DEM/VAL and EMD phase followed by a production phase for a military TAV program.

2. RDT&E cost is assumed to include all the nonrecurring cost to perform risk reduction demonstrations and design trades, along with those recurring activities required to produce, fully test, and demonstrate one subscale "X" demonstrator vehicle and one operational "Y" protoflight vehicle.[1] A production phase would follow this combined phase with the manufacture and delivery of operational military TAVs.

3. All costs for the RDT&E and overall LCC assessments are estimated in base year FY 1997 dollars (this was the earliest first year that an assumed combined DEM/VAL and EMD phase could feasibly begin).

4. Both the D-57 and NK-31 engines that are defined as part of this cost baseline are assumed to go through a maximum of 200 test firings before being flight qualified.

5. As part of the multimission capability described earlier in this report, it is assumed that the payload bay will be designed to handle a containerized unit or units that fit within the overall payload volume, weight, and power margins that will be specified for the military TAV.

6. Several cost elements are **not** included as part of these cost assessments.

   a. No additional cost is estimated for the design, modification, and procurement of either the Boeing 747 air-launcher carrier or the KC-135 aerial-refuel tanker required as the first stage for the launching the "X"

---

[1]The initial cost assessment using Black Horse described in Appendix A assumed erroneously that only an "X" vehicle is required prior to production.

and "Y" vehicles for the two concepts. This same ground rule applies to the production phase of the operational vehicles as well. Actual costs depend on the availability and cost of these aircraft for DoD use and the extent of actual modifications required.

For the Boeing 747, it is assumed that the aircraft used to transport the space shuttle from Edwards AFB to Kennedy Space Center has sufficient load-bearing elements to handle the launch of a sub-scaled version. The KC-135 has performed refueling missions in the past. The only issue is the extent of modification, if any, required to transfer oxidizer to a military TAV. For the purposes of this cost assessment, these design determinations and related cost estimates are not included.

b.   As mentioned earlier, no additional mission-related cost is included for the design, development, procurement, integration, and testing of the payloads required for the military TAV "X" and "Y" vehicles. This same ground rule applies for the production phase of the operational vehicles as well. The size of the payload bay is known, but the type and mix of possible payloads are unknown at this time.

c.   It is assumed that military TAVs will be inherently designed to handle survivability requirements primarily by having significant cross-range or maneuverability capability. Consequently, no added RDT&E and procurement cost is included to provide additional survivability treatments through either shielding of exposed structural portions of the vehicle or electromagnetic pulse testing of critical electronic components.

d.   The RDT&E and recurring production costs do not include government and program office contractor support to perform the program management and system engineering technical activities. The costs displayed are total contractor costs comprised of government-reimbursed and internal contractor funds.

7. It is assumed that RDT&E and production costs reflect timely and adequate government funding typical of multiyear procurements.

8. The contractor RDT&E and production cost estimates presented in this report represent the contractor's new ways of doing business (e.g., "skunk works" type of environment). The key characteristics of this type of operation are

- reduced dedicated contractor team,
- simplified drawing release system,
- no duplication of inspections,

- minimal documentation required, and

- stable customer requirements.

9. For operating and support cost purposes, each operational military TAV will be designed to support a service life consisting of on the average ten launches per year over a ten-year service life for a total of 100 flights. Recurring costs are estimated for one operational military TAV and also for an additional procurement of a fleet of six.

10. A 95 percent cost improvement curve is assumed over the operational launch vehicle production quantity of six.

RDT&E Cost Assessments

Cost Model Evaluations and Advantages of TRANSCOST

Several launch vehicle cost models (LVCMs) and aircraft estimating approaches were evaluated before selecting and adapting the TRANSCOST 6.0 (*Statistical-Analytical Model for Cost Estimation and Economical Optimization of Space Transportation Systems*)[2] as the primary "tool" for estimating reusable military TAVs. Specific TRANSCOST cost estimating relationships (CERs) were modified to adjust for the contractor's "new ways of doing business" as described in the eighth ground rule and assumption listed above and as identified on Figure B.2 in Appendix B as part of a sample output report from the RAND military TAV cost model. The TRANSCOST model CERs were also modified as described later in this section to reflect the current technical maturity assessments of the military TAVs under evaluation compared to those vehicles and subsystems used as part of the TRANSCOST data base.

The NASA Marshall Space Flight Center LVCM developed in 1983 by the Planning Research Corporation[3] was evaluated along with an updated LVCM version developed in 1993 by Tecolote Research[4] for the Air Force Cost Analysis Agency and Army Space and Strategic Defense Command. Both models focus on ELVs and missile cost data with a lower-level mix of solid and liquid rocket engines, avionics, and tanks.

---

[2]Dietrich E. Koelle, TRANSCOST 6.0 (*Statistical-Analytical Model for Cost Estimation and Economical Optimization of Space Transportation Systems*) , 1995 Edition (version 6.0), TCS-TR-140(95), July 1995.

[3]*NASA Marshall Space Flight Center (MSFC) Launch Vehicle Cost Model (LVCM)*, Planning Research Corporation, prepared for NASA MSFC, 1983.

[4]Peter C. Frederic and Arve R. Sjovold, *Launch Vehicle Cost Model Update*, Tecolote Research Corporation, Cost Driver Identification Task 93-013, prepared for U.S. Army Space and Strategic Defense Command and U.S. Air Force Cost Analysis Agency, 16 August 1993.

20

Although the updated Air Force LVCM had planned to incorporate an approach for reusability, the space shuttle is the only U.S. data point that is available and the shuttle design uses technology older than what the RLV concepts are proposing to use. From the most recent information received, the Air Force LVCM database is not stratified sufficiently to develop separate CERs for manned versus unmanned platforms and for metallic versus nonmetallic structures. In addition, the liquid rocket engine data considered only one propellant mix of LOX/LH2 (liquid oxygen/liquid hydrogen) for main engines only with fixed cycle times. The solid rocket engine is based on a more extensive and varied historical database.

Traditional aircraft cost models such as an airframe model developed by RAND in 1987[5] derive the production cost of the first ten airframes as a function of aircraft maximum speed and empty weight (i.e., with no fuel, ordnance, or crew aboard). The CER is based on data from post-1960 aircraft and an average empty or dry weight of 48,000 lb. Applying this CER to exotic structural materials required for RLVs introduces uncertainty into the estimates. In addition, the speed regime of RLVs is well beyond even the supersonic values associated with the aircraft in the database. Any attempt to extrapolate costs on this basis results in adding to the overall cost uncertainty. Also, the remaining subsystem estimates are derived on a percentage basis of this-airframe-only cost, increasing further the uncertainty of the total estimated costs. Finally, even if commercial practices are assumed, space-qualified components for this TAV will most likely be designed and tested to meet a more stringent environment than airborne components.

The advantage of using TRANSCOST is that the model addresses cost differences between ELVs and RLVs on a technical basis using the best historical data available for developing CERs. In addition, this cost model reflects:

1. Development CER differences between conventional medium or low chamber pressure expendable engines and advanced technology high chamber pressure reusable engines as a function of technical quality, reliability, and the number of test firings comparisons.

    Figure 4.1 illustrates the RDT&E CERs from the TRANSCOST 6.0 cost model for engines as a function of mass or weight. Note that both scales are log-log with RDT&E efforts (and therefore costs) for liquid rocket engines being consistently higher over the broad weight range than solid rocket motors.

---

[5]R. W. Hess and H. P. Romanoff, *Aircraft Airframe Cost Estimating Relationships: Study Approach and Conclusions*, RAND, R-3255-AF, December 1987, and updated CERs for advanced materials and composites in S. A. Resetar, J. C. Rogers, and R. W. Hess, *Advanced Airframe Structural Materials— A Primer and Cost Estimating Methodology*, RAND, R-4016-AF, 1991.

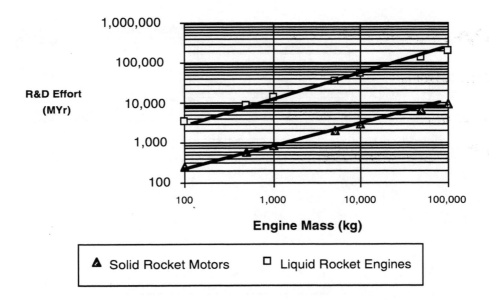

Figure 4.1— Sample TRANSCOST 6.0 RDT&E Engine CERs

This increase in effort is relatively constant for the engines over a fairly wide weight range.

2. Vehicle CERs where dry weight input parameter estimates of RLVs are projected to be 50 percent higher than comparable ELVs because of higher safety factors in the structural design due to the reusability and repeated load cycles, technical performance measurements (TPMs) required for reentry, additional equipment for integrated checkout or health control systems, and increases in redundancy. The higher the vehicle dry weights, the higher the propellant weight for the same size payload, which again increases the overall vehicle weight. This difference between ELVs and RLVs is also reflected in other CERs.

Figure 4.2 displays the RDT&E vehicle CERs for both RLVs and ELVs as a function of mass or weight when the weight is dry weight and excludes the engine weight. All "mass" terms or weights used in this cost assessment are assumed to be dry weights. Note that the RDT&E effort (and cost) difference between RLVs and ELVs is consistently higher over a fairly broad weight range, but the difference declines significantly as the vehicle dry weight of RLVs and ELVs increases.

3. A CER is defined for refurbishment or overhaul costs to reestablish flight readiness separate from those standard maintenance activities required between flights. Refurbishment costs could occur, for example, after every

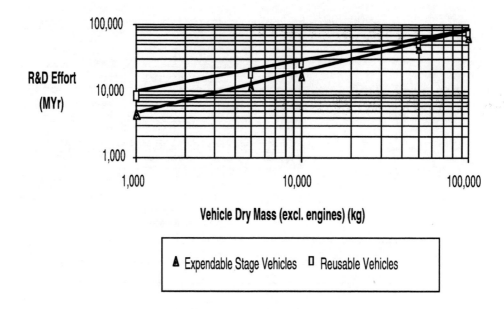

Figure 4.2 — Sample TRANSCOST 6.0 RDT&E Vehicle CERs (Excluding Engines)

fifth flight to include engine overhauls and other preventive maintenance activities that require removal and replacement of components soon to wear out. This CER can also be adjusted to accommodate changes in the frequency of refurbishment over the RLV's service life.

4. CERs are defined for direct operations costs (DOC), which include:

- RLV mission operations for a mission control center and extended global communications;
- standard maintenance activities of RLVs between flights that reflect the much longer and more demanding launch and mission activities than comparable ELVs;
- launch fees and insurance to account for both ELV and RLV potential catastrophic failures and mission aborts with different vehicle reliability, failure, and abort rates depending upon the vehicle being estimated;
- amortized recurring unit production costs of the RLVs over the expected number of potential flights; and
- ground transportation and recovery of RLVs from a remote landing site back to the original launch site.

Technology Readiness Levels and Risk Assessment Adjustments

The TRANSCOST model-generated RDT&E nonrecurring CER-generated estimates are adjusted based on an assessment of the technology readiness levels (TRLs) of the major subsystems and associated risk levels. Figure 4.3 provides the standard NASA definitions for each TRL[6] and the associated qualitative risk assessment levels. Estimated subsystem level costs are scaled up or down after comparing and normalizing the risk assessment levels with the programs and associated risk levels the TRANSCOST model used as the basis of the CERs. TRANSCOST model-generated costs are then adjusted using a linear multiplier to denote the risk level differences between the proposed concept and what is reflected in the database.

TRL	Definition	Risk Level
9	"Flight proven" through successful mission operations	Low
8	"Flight qualified" through ground or flight test and demonstration	
7	Prototype demonstrated in a flight/space environment	Low
6	Prototype demonstrated in a relevant ground or space environment	Low
5	Component and/or breadboard validated in a relevant environment	Moderate
4	Component and/or breadboard validated in a laboratory environment	Moderate
3	Analytical and experimental critical function and/or characteristic proof-of-concept demonstrated	High
2	Technology concept and/or application formulated	High
1	Basic principles observed and reported	High

Figure 4.3—Technology Readiness Level (TRL) Definitions

The technology readiness levels and associated risk levels were assessed across the following subsystems for the representative TSTO military TAV cost baseline concepts defined in Section 3. The technical cost baseline description assumed for the concepts is also provided.

---

[6]These TRLs were first defined as one chart in an overall risk assessment briefing on "Technology Readiness Levels" by NASA Goddard Space Flight Center, 1988. A similar rating scale was applied and used a few years later on the Air Force Brilliant Eyes program. In addition, there is significant evidence that all these risk rating scales may have been based on the earlier work of Mr. Frederic D. Maxwell, Aerospace Corporation, who developed a risk driver assessment framework in the mid-1980s.

- Propulsion (primarily liquid rocket engines)
- Structures and materials
    - Airframe (e.g., titanium metal matrix, graphite epoxy, carbon-carbon)
    - Propellant tanks (e.g., aluminum/lithium, graphite epoxy)
    - Engine structures (e.g., graphite/polyimide, narloy/zenon/ honeycomb panels)
- Thermal protection system (e.g., advanced carbon-carbon, refractory composite insulation tiles, carbon/silicon carbide, flexible blanket insulation)
- Aerosciences (includes hypersonic aerodynamics, controllability, ascent/reentry shock and loads, heating effects)
- Avionics (including vehicle health management and monitoring, engine monitoring, thermal management, power source, data bus, sensors)
- Operations (e.g., production, delivery, and maintenance of propellants; environmental management for ground operations; ground-based engine health management and monitoring, etc.).

The corresponding assessed risk levels are provided in Table 4.1 for both TSTO versions.[7] Except for some high-risk-level concerns in the aerosciences area, most risks are either moderate or low for the two military TAV cost baseline concepts.

Table 4.1

Military TAV Quantitative TRL Assessments

Area	Assessed Risk Level
Propulsion	Low
Airframe	Low to moderate
Propellant tanks	Low to moderate
Engine structures	Low to moderate
Thermal protection system	Moderate
Aerosciences	Moderate to high
Avionics	Moderate
Operations	Low

Cost Estimating Cross Checks

To minimize the uncertainty in the cost estimates from our modified use of the TRANSCOST CERs and improve the confidence level in the estimates, some of the other cost models and estimating approaches noted above have been used as

---

[7]See MR-890-AF, Section 4, for a further description of these technical subsystems and associated risks.

secondary "tools" to cross check against our output. An example of cross checking is to use the (primarily) ELV costs derived from the Air Force LVCM as a lower bound for military TAV costs and apply shuttle-analogous estimates as an upper bound. Not only do these cross checks help to reduce uncertainty and improve the confidence in the final estimates, but the final estimates can be more accurately articulated as upper bound ("not to exceed") budgetary-type values.

RDT&E Cost Results

RDT&E cost assessments include estimates of both TSTO military TAV representative cost baseline concepts. In addition, we compared launch vehicle dry weights, development complexity, and estimated costs of an existing ELV, Pegasus, to the two concepts. All three vehicles are potentially capable of handling at least some of the payload range requirements envisioned for a military TAV, although each of the vehicles may not satisfy all the projected military operational needs. A summary of the RDT&E cost assessments is provided in Table 4.2. The cost model report details may be seen in Appendix B.

The Pegasus RDT&E cost results are presented as a commercial ELV data point comparable with the military TAVs. Of all the operational ELVs, Pegasus comes closest to handling the payload requirements and timeliness of a military TAV. Pegasus can handle a 800 lb payload at 100 nmi at 90 degrees inclination.[8] The

Table 4.2

Military TAV RDT&E Cost Estimates (FY97 $M)

Factor	Pegasus Expendable	TSTO Aerial-Refueled TAV	TSTO Air-Launched TAV
Total vehicle dry weight (lb)[a]	6,615	25,000	34,240
Payload weight to LEO (lb)	800	1,600 - 2,000[a]	3,000 - 6,000[a]
Total engine weight (lb)		4,535	3,000
		(D-57, NK-35)	(D-57)
Engine RDT&E cost ($M) [b, c]		129.0	87.0
Vehicle RDT&E cost ($M) [b, d]		630.0	590.0
Total RDT&E cost ($M)	149.0	759.0	677.0

[a]TAV dry weights assume an unmanned version for both TSTO concepts. The Pegasus dry weights are computed by subtracting gross mass less propellant mass from the three launch vehicle stages.

[b]Costs are based on modified use of TRANSCOST 6.0, 1995 Edition (version 6.0) CERs.

[c]Rocket engine R&D costs are based on engine weight and an assumed 200 test firings required for flight certification and qualification.

[d]Vehicle R&D costs are based on vehicle dry weight (excluding engines), design maturity, and team experience.

---

[8]Steven J. Isakowitz, *International Reference Guide to Space Launch Systems*, (Second Edition), American Institute of Aeronautical Engineers (AIAA), 1991. All Pegasus weights and program schedules were extracted from this document.

Pegasus RDT&E cost estimate of $149M is based on a modified use of TRANSCOST 6.0 CERs similar to both TSTO military TAV versions. This $149M estimate compares with the TSTO air-launched military TAV RDT&E cost estimate of $677.0M. Finally, adding the development and test costs of an additional engine, the RDT&E cost for the TSTO aerial-refueled concept is estimated to be $759.0M (12 percent higher than the air-launched version).

Overall LCC Assessments

LCC Methodology

Modified TRANSCOST CERs are used to derive recurring production and O&S costs. The estimating approach generates LCC for the two TSTO military TAV concepts. The TAV concept with the lowest LCC is identified as the air-launched version and treated as the upper-bound budgetary estimate, where one can achieve the full use or value of the vehicle over the assumed ten-year service life. Another way to express full use or value is that this budgetary estimate represents the total ownership cost associated with using this TAV TSTO version for ten years with an average of ten launches per year or 100 total launches. The purpose of this assessment is to represent how many more launches or sorties military TAVs as reusable vehicles can be achieved over ELVs given the same available budget. The assessment is based on the assumption that the lowest LCC estimate of the three vehicles—the air-launched version—is the same budget available for the other two vehicles. The results and explanations of the computations for this assessment are described below.

LCC Affordability Assessment Results

First, LCC were estimated for each of the three alternatives; the vehicle identified with the lowest cost was the TSTO air-launched TAV—$1,862M,[9] which represented the budget ceiling for this vehicle and all the other alternatives. See Table 4.3.

The number of equivalent launches of the aerial-refueled TSTO version is then calculated by starting with the budgetary estimate of $1,862M and (1) subtracting out the RDT&E estimated cost of $759M and the $60M for the unit production cost for one vehicle, and then (2) dividing the remaining cost of $1,043M by the

---

[9]The total LCC estimate for the TSTO air-launched version is computed by adding (1) the RDT&E cost estimate of $677.0M, (2) the unit production cost for one vehicle of $55.0M, and (3) the O&S cost per launch of $11.3M times the 100 launches.

Table 4.3

Military TAV Life-Cycle Cost Estimates (FY97 $M)

Factor	Pegasus Expendable	TSTO Aerial-Refueled TAV	TSTO Air-Launched TAV
Total RDT&E cost[a]	$149.0	$759.0	$677.0
Avg. recurring unit production cost[a]	$17.0	$60.0	$55.0
Avg. refurbishment cost/launch[a]	N/A	$1.7	$1.5
Direct operations cost/launch[a]	$4.7	$7.6	$6.1
Indirect operations cost/launch[a]	$3.1	$3.9	$3.7
Launch insurance/launch[a]	$1.9	(included in DOC)	(Included in DOC)
Total recurring cost/launch	$26.7[b]	$13.6[c]	$11.3[c]
Total LCC TSTO air-launched budget for 1 vehicle	$1,862.0	← $1,862.0	← $1,862.0
Equivalent number of launches	64	79	100
Total LCC TSTO air-launched budget for 6 vehicles	$7,787.0	← $7,787.0	← $7,787.0
Equivalent number of launches	286	505	600

[a]All costs are computed based on modified use of CERs from the TRANSCOST cost model, 1995 Edition (version 6.0).

[b]The total recurring cost per launch for the expendable Pegasus vehicle includes the unit procurement cost of $17.0M and the recurring O&S cost per launch of $9.7M (adding up the $4.7M, $3.1M, and $1.9M estimates listed in the table).

[c]The total recurring cost per launch for two TAV versions includes only the total recurring O&S costs from the three cost elements listed within each column.

recurring estimated O&S cost per launch of $13.2M. These calculations result in a budget-equivalent TSTO aerial-refueled version of 79 launches (rounded down to the nearest whole integer).

The number of launches is also computed for the Pegasus ELV using this same TSTO air-launched budget estimate and (1) subtracting out the total RDT&E estimated cost of $149M and then (2) dividing the remaining cost of $1,713M by the total recurring cost per launch of $26.7M, which is the total of the recurring expendable unit production cost of $17M and the recurring estimated O&S costs per launch of $9.7M. These calculations result in a budget-equivalent computed number of expendable Pegasus launches of 64 (rounded down to the nearest whole integer).

The set of computations is repeated given a fleet of six vehicles,[10] with six being the multiplier for the $55M TSTO air-launched version and $60M for the aerial-refueled version. The number of equivalent launches is again compared

[10]For a fleet of six vehicles, it is assumed that no additional RDT&E costs are required for the additional buy and the average recurring unit production cost is the same as the estimates used for the one-vehicle calculations.

assuming that the lowest LCC estimate of $7,787.0M for the TSTO air-launched TAV concept represents the same total maximum budget available for the other two alternatives.

LCC Affordability Observations

The economic benefit of reuse is apparent whether one or a fleet of six military TAVs is procured. An equivalent number of aerial-refueled TSTO TAVs and Pegasus ELV launches are compared with the same budget.

With a $1.9B budget for one vehicle, the TSTO air-launched military TAV can provide over 26 percent more launches than the TSTO aerial-refueled version. In addition, this air-launched TAV version can provide over 56 percent more launches over the ten-year service life than the Pegasus ELVs given the same budget.

Given a $7.8B LCC budget, a fleet of six TSTO air-launched military TAVs can provide over 18 percent more launches than the TSTO aerial-refueled version with the same number of reusable vehicles. In other words, for the same amount of money, a fleet of six aerial-refueled TAVs can provide only 84 percent of the number of launches possible with a fleet of six air-launched vehicles.[11]

Figure 4.4 expresses the results of Table 4.3 graphically by comparing the LCC per launch over the payload lift capability to LEO of the two TSTO vehicle concepts and the Pegasus ELV. The cost per pound for each vehicle is displayed as a data point.

The launch costs for all three vehicles shown in Figure 4.4 represent the cost to the user. Several launch costs are not included:

- The RDT&E and procurement cost of the payload itself and the recurring cost of the assembly, integration and testing (AI&T) of the payload with the particular launch vehicle.[12]

---

[11]The 84 percent average utilization is based on dividing the total number of equivalent launches of 505 for aerial-refueled TAVs by the six reusable vehicles. The percentage utilization is based on the ratio of the resulting number of flights per vehicle of 84 (rounded off) over the 100 average flights possible over the service life.

[12]Even though the cost of the payload AI&T may be slightly different for each alternative the payload costs are the same, so the overall observations to be drawn from this figure should still be valid.

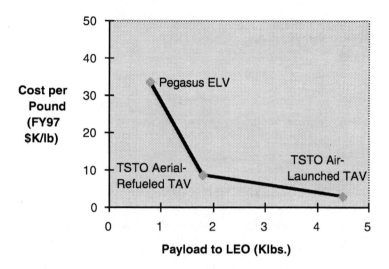

Figure 4.4—Economic Advantages of Military TAVs over ELVs

- The government costs, including the costs of the first-stage aircraft required for the vehicle to launch (or refuel from) before ascending into a suborbital or orbital trajectory.[13]

However, even when these costs are added, the overall economic advantages of reusable over expendable vehicles should remain.

In summary, the military air-launched TSTO TAV has significant economic reuse value, in terms of the increased number of expected launches, utilization rates and cost per pound of payload lift, over both the aerial-refueled version and the Pegasus ELV. Furthermore, if comparable expendable vehicles are not always available for use when requested, the differences in the number of launches of fully utilized reusable vehicles over Pegasus could end up being even larger than the 600 versus 286 launches indicated in the table above.

---

[13]There may be additional minimal costs per pound required for the air-launched version for RDT&E budget for modifying the first-stage carrier aircraft for extra load stability and propulsion lift capability and for covering additional recurring O&S and fuel consumption costs of the carrier aircraft. For the aerial-refueled version, there may be additional RDT&E costs that will need to be amortized that are associated with in-flight refueling and the special propellant required for reusable vehicles. In addition, there may be additional recurring O&S and fuel consumption costs associated with using this tanker aircraft for aerial refueling.

5. Observations

Before discussing potential implementation strategies for a military TAV development program, we briefly review the results of our military TAV cost assessment.

Summary of Cost Assessments

The independent cost assessment performed by RAND indicates that it should be possible to develop an air-launched military TAV for under $700M and that such a vehicle would be capable of delivering up to 5,000 lb of payload into LEO. This type of military TAV (the orbital rocket-powered vehicle and not the first-stage carrier aircraft) would be much smaller than a commercial RLV. The RLVs envisioned by NASA and industry would be designed to serve customers of the current commercial launch market and those companies and nations that need medium-sized satellites delivered to geostationary orbit. Development costs for a commercial RLV sized to the primary commercial satellite launch market vary from $3B to $20B for a single vehicle.[1]

Our analysis indicates that a fleet of six military TAVs can provide over twice as many launches as the Pegasus expendable vehicle for the same overall LCC budget and deliver over six times as much total payload weight to orbit over the same life cycle. Furthermore, the analysis indicates that it may cost an order of magnitude less to deliver a pound of payload to LEO using a small military-type TAV, such as the TSTO NGC TAV, than an expendable launch vehicle like Pegasus (with launch costs of $2900/lb as opposed to $34,000/lb). This cost difference would represent a significant competitive advantage for a commercial launch services provider.

A major programmatic decision that drives RDT&E costs in our analysis is the assumption that the risk reduction portion of the overall development of a military TAV could be under way now. However, NASA and Lockheed Martin, the X-33 prime contractor, are developing enabling technologies that will mature in the FY 1999 time frame. DoD and the Air Force could take advantage of these technologies by initiating risk reduction only where needed and delaying the

---

[1]Joseph Anseluno, "NASA Nears X-33 Pick," *Aviation Week and Space Technology*, June 17, 1996, p. 24.

start of military TAV EMD-type activities. If this were done, total RDT&E costs for a military TAV could potentially be lower than the $700M figure estimated above.

Potential Implementation Strategies

Because of the extremely low launch costs (as measured by cost per pound of payload delivered to orbit) that might be achieved using a military TAV, a small TAV could provide competitive launch services in commercial small-satellite markets. In other words, a small TAV designed to satisfy military user needs would be a dual-use system that could also be used in the commercial marketplace.

In the near future, commercial satellite developers and customers may make greater use of small satellites for remote sensing as well as communications. In the next few years, a number of commercial communication satellite constellations will be deployed in LEO to serve emerging markets for cellular phone and internet access services. Examples of such systems include Iridium, a constellation of 66 small satellites, and Teledesic, a constellation of perhaps up to 900 small satellites. These emerging systems represent new business opportunities for small-satellite launch service providers and may encourage significant new private investment in development of small TAVs.

Consequently, it may be possible for DoD and the Air Force to share with industry the investment of RDT&E and vehicle production costs for a small TAV, if the program can be structured appropriately. It may even be possible for the government to share some operations and maintenance costs with industry. For example, if a TSTO TAV system were developed, the first-stage carrier aircraft could be shared or loaned to the Air Force on a basis similar to CRAF (Civil Reserve Air Fleet) during crises or war to meet surge requirements.

One example of an innovative financial structure for such a program is the NASA-industry cooperative agreement notice for the development of the X-33 sub-scale prototype vehicle, although a different financial model may be more appropriate in the small TAV case.

A possible cost-sharing arrangement is for DoD to underwrite early RDT&E investment and then to share in vehicle production start-up investment and capitalization costs. The first few TAVs to come off the production line would be delivered to the Air Force and their capabilities proven in military operations. Thus, the Air Force would make some sort of up-front commitment to buy and own a small quantity of TAVs from the first-block build of vehicles to meet DoD peacetime readiness and core wartime needs. Later TAVs from the production line would go to the commercial launch service provider.

The guarantee of government revenue for procuring TAVs early in production provides for a higher probability that corporate internal rate of return estimates and other financial goals can be met. By ensuring that corporate financial goals can be met, including the government-underwritten guarantee of sufficient cash flow, the TAV producer can keep the production line open and produce additional TAVs for commercial customers. The number of additional TAVs built would be a function of the commercial market for small-satellite launches that can be captured from ELV suppliers at the time of entry into the marketplace. As the market matures and expands, the launch service industry and the fleet of small commercial TAVs can be expanded to meet market demand.

In addition, if during crisis or war a military surge capability is needed, DoD could take advantage of these commercial TAVs (assuming DoD payload commonality requirements are designed into the original production) to launch additional satellites on demand. Some commercial satellite launches could be rescheduled to allow for the timely launch of DoD supplementary space assets.

Regardless of what type of cost-sharing arrangement is made, a "win–win" acquisition strategy for the Air Force and for industry could be implemented if an innovative acquisition approach is adopted.

Appendix

A. Initial Military TAV Affordability Assessment

Introduction

Two major differences are reflected in the initial cost assessment described in this report. The estimated costs for the assessment are based on analogous dollars per pound and complexity differences among Black Horse, Pegasus, and the original NASA X-33 vehicle and the TRANSCOST model. RDT&E costs for the assessment include only a technology demonstrator or X vehicle and not the X and Y vehicles.

Initial RDT&E Cost Assessment

Table A.1 provides an initial affordability assessment that identifies an upper-bound estimate of the total RDT&E budget required to complete an EMD phase for a military TAV program, including the delivery of one technology demonstrator. Development cost is assumed to include all the nonrecurring costs to perform the design trades described in the report along with those activities required to produce a technically feasible operational vehicle design. A production phase would follow EMD with the manufacture and delivery of operational military TAVs.

This initial affordability assessment is based on a comparison and relative sizing of dry weights, costs, and development complexity differences of three similar launch vehicles: Pegasus, X-34, and the TSTO Black Horse; these vehicles represent an existing system, a technology demonstrator, and a conceptual design, respectively.[1] Each vehicle is capable of potentially handling at least some of the payload range requirements envisioned for a military TAV, although the vehicles may not satisfy all the military mission and operational needs. Each vehicle and the supporting data used in this affordability assessment are described in more detail below.

---

[1] All weights are assumed to be dry weights. Development complexities are values assigned to each launch vehicle that are used as linear multipliers to provide a cost scale that measures the relative technical, mission, and operational differences between the three launch vehicles.

All costs in the initial military TAV RDT&E Affordability Assessment are displayed in constant fiscal 1997 dollars. Using the TSTO Black Horse as a representative basis for a military TAV, an upper bound of $500M in constant fiscal 1997 dollars is displayed as the representative required upper bound RDT&E budget.

As seen in Table A.1, the initial estimated cost of developing and testing a military TAV Black Horse demonstrator is greater than the comparable estimated RDT&E costs of Pegasus at $149.0M and the original NASA X-34 program budget of $170M. However, this initial military TAV RDT&E estimate of $500M, even escalated to "then year" dollars, is considerably less than the Phase I contractors' comparable total development cost estimate for the NASA X-33 program of $3B to $10B required through the turn of the century. It is our belief that this upper-bound initial DoD RDT&E budgetary projection maybe a reasonable investment toward achieving the responsiveness and flexibility required to meet the military mission and operational needs described in the report. An initial overall life cycle cost (LCC) assessment of the military TAV that includes this initial RDT&E cost and recurring production, operations, and maintenance cost is provided at the end of this appendix.

Table A.1

Initial Military TAV RDT&E Affordability Assessment

Factor	Pegasus	X-34	Black Horse
Total dry weight (lb)	6,615	11,200	16,398
Payload weight capacity (lb)	800	1,200	1,000
Development complexity	1.0	1.4	2.8
Expendable RDT&E cost ($/dry lb)	$25,600	N/A	N/A
Reusable RDT&E cost ($/dry lb)	N/A	$15,180	$30.490
Total RDT&E cost (FY 97 $M)	$149.0	$ 170	$500[a]

SOURCES: All Pegasus weights and program schedules were extracted from the AIAA *International Reference Guide to Space Launch Vehicles*, 1991 edition. All X-34 weights were extracted from "X-34 Program Overview Presented to the TAV Workshop for Project AIR FORCE RAND, by Dr. Antonio L. Elias of Orbital Sciences Corporation, April 1995. All X-34 costs and program schedules are extracted from "Industry Steps Up to Reusable Rocket Effort—OSC, Rockwell Selected to Run X-34 Project," *Space News*, 13–19 March 1995, pp. 4, 37. All Black Horse weights were extracted from the "Aerial Propellant Transfer TAV" briefing to RAND by Capt. Mitchell Burnside Clapp, Air Force Phillips Laboratory, February 1995.

[a]The Black Horse computed dry weight is used along with the computed reusable dollars per dry weight of the original X-34 vehicle times the development complexity of 2.8 to estimate an upper-bound budgetary ROM estimate of $500M in constant fiscal 1997 dollars.

Initial Launch Vehicle Details

Pegasus is the commercial expendable-vehicle data point for projecting military TAV RDT&E costs. Of all the operational ELVs, Pegasus comes closest to being capable of handling the payload requirements and timeliness of a military TAV. Pegasus can handle a 800 lb payload at 100 nmi at 90 degrees inclination. [2] Total development cost of Pegasus is estimated at $149.0M using the same cost methodology as Black Horse. These costs are then used as the basis for computing the dollars per pound of dry weight for Pegasus. Pegasus is assigned a development complexity value of 1.0.

Next, the multiple-stage-to-orbit reusable X-34 vehicle represents a civil/commercial data point for projecting military TAV development costs on a dollars per pound of dry weight basis equivalent to Pegasus. Of the two NASA Orbital Sciences Corporation (OSC) original X-34 concepts, the Lockheed L-1011-launched reusable configuration represents a design with a dry weight capable of carrying a 1,200 lb orbital payload at the same altitude and orbital inclination as Pegasus.[3] The original X-34 RDT&E budget is $70M from NASA plus an initial investment of $50M each from OSC and Rockwell for a total initial program budget of $170M in then-year dollars.[4] The program began in March 1995 with a first launch planned for 1998. Therefore, the same approximate then-year value of $170M is used to approximate fiscal 1997 dollars.[5]

The initial X-34 development complexity is computed from the ratio of computed reusable development cost per lb for X-34 to the computed expendable development cost per lb for Pegasus. This complexity is computed as approximately 1.4, which assumes that the X-34 design is 40 percent more complex than Pegasus after normalizing for cost and dry weight.

Finally, the TSTO Black Horse is assumed to be an initial concept that might meet military TAV mission and operational needs. Black Horse is designed to handle

---

[2]All Pegasus weights, costs, and program schedules were extracted from the AIAA *International Reference Guide to Space Launch Vehicles*, 1991 edition. The Pegasus dry weights are computed by subtracting gross mass less propellant mass from the three launch vehicle stages.

[3]All original X-34 weights were extracted from the "X-34 Program Overview Presented to the TAV Workshop for Project AIR FORCE RAND," by Dr. Antonio L. Elias of Orbital Sciences Corporation, 19 April 1995.

[4]All original X-34 costs and program schedules are extracted from "Industry Steps Up to Reusable Rocket Effort—OSC, Rockwell Selected to Run X-34 Project," *Space News*, 13–19 March 1995, pp. 4 , 37.

[5]The original X-34 program mid-point (where 50 percent of the budget is expended) is assumed to occur in fiscal 1997.

a 1,000 lb payload.[6] The development complexity for the Black Horse design is assumed to be twice that for the X-34—an assumption made to establish an initial upper-bound budgetary estimate based upon the different mission and operational needs and more demanding design of the military TAV over the civil/commercial X-34 launch vehicle. In addition, the doubled complexity (from 1.4 to 2.8) accounts for the design uncertainty in this early Black Horse concept and other technical maturation and associated risk issues. The difference of 1.4 between these two complexities can be interpreted as the military TAV being 140 percent more complex than the X-34 after normalizing for cost and dry weight.

Initial Life-Cycle Cost Assessment

The higher military TAV RDT&E front-end budget of $500M (see Table A.1) should result in lower total recurring cost of a reusable military TAV over an expendable Pegasus. The recurring cost of launching an expendable Pegasus includes the total production cost of each vehicle plus the fixed and variable launch operations cost. The total estimated RDT&E cost for Pegasus of $149M can be amortized over the expected number of operational expendable vehicles procured.

Like Pegasus, the military TAV recurring cost is comprised of fixed and variable launch operations cost. The higher nonrecurring RDT&E budget of $500M can be amortized over the total number of flights anticipated by all the reusable vehicles procured. Because the TAV is reused, the military TAV RDT&E budget can be amortized over a larger number of total flights than can Pegasus with only one operational vehicle per flight. Also, the recurring cost of procuring each operational military TAV can be amortized over the anticipated number of launches per vehicle. The total recurring cost per military TAV launch is comprised of these two amortized costs along with the total launch operations cost per vehicle.

A comparison with Pegasus' recurring unit production and operations cost is used to derive a comparable set of costs for the Black Horse military TAV example. For production, the same recurring unit cost per pound of dry weight for Pegasus is assumed for Black Horse. According to Dr. Antonio Elias of OSC, the current recurring unit cost of Pegasus is around $10.2M per vehicle.[7] If we

---

[6]All Black Horse weights were extracted from the "Aerial Propellant Transfer Space Plane" briefing to RAND by Capt. Mitchell Burnside Clapp, Air Force Phillips Laboratory, February 1995. For Black Horse, total dry weight is computed from takeoff gross weight less propellant weight.

[7]The recurring unit production cost of expendable hardware for a fully reusable three-stage launch vehicle of $10,207,420 was extracted from the "X-34 Program Overview Presented to the TAV Workshop for Project AIR FORCE RAND," by Dr. Antonio L. Elias of Orbital Sciences Corporation,

apply the same computed Pegasus recurring unit cost per pound of dry weight to the military TAV Black Horse dry weight, the recurring unit production cost of the military TAV would be $26.8M per vehicle in constant fiscal 1997 dollars.[8]

For recurring operations costs, it is assumed a military TAV will have fixed and variable operations costs comparable to Pegasus. Even though not stated, we assume Pegasus operations costs include some variable maintenance cost to repair the vehicle during pre-launch, checkout, and inspection, and test as needed. A military TAV will require comparable, but probably higher, variable maintenance cost between flights. These costs will vary as the overall reliability of the reusable vehicle changes over time. In addition, a reusable vehicle will incur fixed maintenance costs to overhaul the engines, airframes, and other components after a specified number of launches. Dr. Elias of OSC stated that the current recurring fixed and variable operations cost per launch for Pegasus is around $1.6M per vehicle in constant fiscal 1997 dollars.[9] As an upper bound, we assume that the combined average recurring operations and maintenance costs per launch for the Black Horse will be double the recurring operations cost for the Pegasus. The recurring cost per launch for the military TAV would be $3.2M per vehicle in constant fiscal 1997 dollars.

A total LCC budget is generated for procuring one reusable vehicle and a fleet of six vehicles. For a relative economic sense of what value a $500M military TAV RDT&E and total LCC budget may provide, an equivalent number of launches is computed for Pegasus given the two Black Horse procurement cases where one military TAV is assumed to have a reliability that allows each vehicle to be reused for a minimum of 100 flights and a fleet of six for 600 flights.

The results are summarized on Table A.2. The same total LCC for one operational military TAV of $846.8M can be amortized across 100 launches. This same total LCC budget covers an equivalent of only 56 launches with a Pegasus,

---

19 April 1995. It is assumed that this cost is representative of the fiscal 1995 estimate for procurement of a Pegasus launch vehicle. Escalation is based on using a nominal compounded inflation factor of 3 percent per year for two years from fiscal 1995 until fiscal 1997, resulting in a Pegasus recurring unit cost of $10.8M in constant fiscal 1997 dollars.

[8] The recurring unit production cost of Pegasus of $10.8M is divided by the dry weight for Pegasus from Table 3.1 of 6,615 lb to result in a dollars/dry weight of $1,637 per lb. This value is then multiplied by the Black Horse dry weight from Table 3.1 of 16,398 lb to derive the recurring unit production cost of Black Horse of $26.8M.

[9] The total recurring fixed ($1,000,000) and variable ($496.068) operations costs for a fully reusable three-stage launch vehicle of $1,496,068 was extracted from the "X-34 Program Overview Presented to the TAV Workshop for Project AIR FORCE RAND," by Dr. Antonio L. Elias of Orbital Sciences Corporation, 19 April 1995. It is assumed that this estimate is representative of the fiscal 1995 cost for operating a Pegasus launch vehicle. Escalation is based on using a nominal compounded inflation factor of 3 percent per year for two years from fiscal 1995 until fiscal 1997, resulting in a Pegasus recurring operations cost of $1.6M in constant fiscal 1997 dollars.

Table A.2

Initial Military TAV Life Cycle Cost Affordability Assessment

Factor	Pegasus	Black Horse
Total RDT&E cost (FY97 $M)	$149.0	$500
Recurring cost ($/dry lb)	$1,637	$1,637
Recurring unit production cost (FY97 $M)	$10.8	$26.8
Recurring operations & maintenance cost per launch (FY97 $M)	$1.6	$3.2
Total LCC budget (one vehicle) (FY97 $M)	N/A	$846.8
Equivalent number of launches (one vehicle)	56	100
Total LCC budget (six vehicles) (FY97 $M)	N/A	$2,580.8
Equivalent number of launches (six vehicles)	196	600

SOURCES: The recurring unit production cost of expendable hardware for a fully reusable three-stage launch vehicle of $10,207,420 was extracted from the *X-34 Program Overview Presented to the TAV Workshop for Project AIR FORCE RAND*, by Dr. Antonio L. Elias of Orbital Sciences Corporation, 19 April 1995. The total recurring fixed ($1,000,000) and variable operations costs ($496,068) for a fully reusable three-stage launch vehicle of $1,496,068 was extracted from the Overview.

given the higher military TAV recurring launch operations costs.[10] For a fleet of six military TAVs and no unit production cost improvements (i.e., no effects of learning) over the procurement quantity, the $500M RDT&E cost combined with the $2,080.8M total recurring unit production and operations and maintenance cost results in a total LCC of $2,580.8M. This LCC budget provides a capability for 600 military TAV launches compared with around 197 launches for Pegasus with the same budget.[11] The equivalent number of launches once again favors Black Horse.

The economic value of reuse is apparent especially when a fleet of six military TAVs is procured and a comparison made of the equivalent number of Pegasus launches computed using the Black Horse LCC budget estimates. Even with an upper-bound estimate for operations and maintenance cost, a fleet of six military TAVs has significant economic reuse value in terms of the expected number of launches over the same budget in expendable Pegasus vehicles.

---

[10]Assuming a total LCC (R&D, production, and operations) budget of $846.8M ($500M + $26.8M +($3.2M × 100 launches)) for Pegasus, $149.0M of the R&D budget has to be subtracted, leaving $697.8M. The $ 697.8M is then divided by the Pegasus recurring cost (production and operations) of $12.4M ($10.8M + $1.6M) to determine the approximate equivalent number of Pegasus launches given this Black Horse budget for one vehicle. This calculation results in 56.3 launches and is rounded to 56 launches for our analysis.

[11]The total recurring production cost for six Black Horse vehicles is assumed to be the unit production cost of $ 26.8M multiplied by six with no learning, or $160.8M. The total operations and maintenance cost for six Black Horse vehicles is assumed to be 600 multiplied by $3.2M, or $1,920M. Assuming a total LCC budget of $2,580.8M ($500M + $160.8M+$1,920M) for Pegasus, $149.0M of the R&D budget has to be subtracted, leaving $2,431.8M. The $2,431.8M is then divided by the Pegasus total recurring cost of $12.4M to determine the approximate equivalent number of Pegasus launches given the Black Horse budget for a fleet of six vehicles. This calculation results in 196.1 launches and is rounded to 196 launches for our analysis.

The TSTO Black Horse concept is used in the affordability assessment as a representative military TAV example. However, this cost analysis can be extrapolated to other military TAV concepts in the same payload class because the complexity values will most likely be similar. Given this same economic value and LCC assessment, DoD could obtain a reusable vehicle that is potentially responsive and flexible enough to meet the military mission and operational needs described in this report.

B. RAND Military TAV Cost Model Sample Output Report

Figures B.1 and B.2 represent a sample output report from the RAND Military TAV cost model based on a modified use of the TRANSCOST 6.0 documented set of CERs. This spreadsheet report reflects results that are summarized on Tables S.1 and S.2 of the Summary and on Tables 4.2 and Tables 4.3 in the main report. Figure B.1 is the input listing and Figure B.2 is the resulting cost-estimating detailed output. Most of the actual CER equations that were used were extracted from the TRANSCOST model report.[1]

---

[1]Dietrich E. Koelle, TRANSCOST 6.0 (Statistical-Analytical Model for Cost Estimation and Economical Optimization of Space Transportation Systems) , 1995 Edition (version 6.0), TCS-TR-140(95), July 1995.

Reference: Modified TRANSCOST 6.0 Cost Model, 1995 Edition, Dietrich E. Koelle			
Input Factors:			
1). FY-97 k$/Man-Year (MY)	$214.2		
	Guidelines	**Inputs**	**Subsystem/Comments**
2). Development Standard Factor (f1)		0.4	Engine & Vehicle
First Generation/new technology	1.2 to 1.3		
New design/some new technology	1.1 to 1.2		
Nom average project, State-of-the-Art	1		
Similar project/no new technology	.7 to .9		
Modification of existing project	.4 to .6		
	Guidelines	**Inputs**	**Subsystem/Comments**
3). Technical Quality Factor (f2)		0.5	Engine
(Engine Development # of test firings linked			(Solid Motor F2 Factor Compared
to reliability.)			to Liquid Engines)
Expendable Liquid 1,000	1		
Reusable Liquid 1,700	1		
500 tests	0.5		RL-10 reusable 20 tests
1000 tests	0.7		performed w/o refurbishment
2000 tests	1.1		
3000 tests	1.2		
	Guidelines	**Inputs**	
4). Team Experience Factor (f3)		0.7	
New Team/No relevant experience	1.3 to 1.4		
Partially new project	1.1 to 1.2		
Company/industry team experience	1		
Team performed on similar projects	.8 to .9		
Team has superior experience	.7 to .8		
	TSTO Aerial-Refueled	**TSTO Air-Launched**	
5). Engine Mass (kg)	2,057.0	1,360.8	
6). Vehicle Dry Mass (excl. engines) (kg.)	9,282.8	14,170.2	
(Reusable dry mass 50% higher than			
expendables for same propellant mass			
due to higher safety factors & repeated			
load cycles, thermal protection for			
re-entry & increased redundancy)			
Net mass increase around 170%			
Net development effort around 200%			
	Guidelines	**Inputs**	
7). Development Schedule vs. Effort		7 yrs.	
8,000 to 15,000 man-year effort	5 yrs.		
15,000 to 30,000 man-year effort	6 yrs.		
30,000 to 50,000 man-year effort	7 yrs.		
70,000 man-year effort	9 yrs.		
90,000 man-year effort	11 yrs.		

Figure B.1—Military TAV Cost Model Report, Input

8). Lockheed Skunkworks Adjustments

	Inputs
Cost Reduction	55%
Schedule Reduction	20%

9). Average Refurbishment Costs for TFU
(Analogy of average % cost/flight
Based on X-15 Plane Data Point)

		Guidelines Per-cent of New Vehicle Recurring Cost	Inputs
	Total	2.60%	2.60%
	Manpower	0.46%	0.46%
	Spares	2.14%	2.14%

This per-cent increases

	Guidelines	Inputs
1-10 flights = First Year	2.60%	2.60%
10-20 flights = 2nd Year	4.46%	4.46%
20-30 flights = 3rd + Years	5.48%	5.48%

10). Pre-Launch Ground Ops Costs
(TSTO with runway take-off
Assumes 10 flights/yr per TAV)

	Guidelines (MY)	Inputs (MY)
Total Effort per year	220-250 Man-Years	220
Vehicle Portion	170-200 Man -Years	
Facilities Portion	50 Man-years	

11). Propellant Costs

	Inputs ($/kg)
Liquid Hydrogen (FY95 $) for Shuttle	$3.25
Assume boil-off at 35%	
Kerosene or Liquid Oxygen (LOX) lower	$0.21

12). Launch & Mission Ops Costs
(includes Mission Control Center and
Extended Global Data Communications)

	Guidelines	Inputs
Vehicle Complexity (D) for each stage		
Unmanned Reusable Orbital Systems	1.4	1.4
Crewed Orbital Vehicles	2.0	

13). Ground Transportation & Recovery

	Inputs Recovery Mass (kg.)
TSTO Air-Launched	15.5
TSTO Aerial-Refueled	11.3
Pegasus	0.0

14).   Fees and Insurance Cost

	Input ($/lb.)
Launch Site User Fee	$2.50
(function of total vehicle dry weight)	

15). RLV Premature Loss Charge Rate

	Inputs
Ballistic SSTO assumes 1 of 500 flights is lost	
Shuttle has experienced 7.5 of 500 flights lost	0.015

16). Surcharge for Mission Abort Rate

SSTO assumes 1 of 30 flights is aborted	0.033
Use a factor of 2 or 3 to include indirect cost (IOC)	

Figure B.1—Continued

	TSTO Air-Launched TAV	TSTO Aerial-Re-Fueled TAV	Pegasus ELV
Non-Recurring Engine Cost (FY-97 k$)	$87,213	$129,313	
Non-Recurring Vehicle Cost (FY-97 k$)	$589,808	$630,053	
Total NRE Cost (FY-97 k$)	$677,021	$759,366	$148,890
Rec TFU Engine Cost (FY-97 k$)	$10,927	$9,120	
Rec TFU Vehicle Cost (FY-97 k$)	$44,106	$51,100	
Total Rec TFU Cost (FY-97 k$)	$55,033	$60,220	$16,980
Average Refurbishment Costs for TFU/flt.	$1,501	$1,737	N/A
Direct Operations Costs (DOC) (50%)	$6,120	$7,592	$4,664
1). Pre-Launch Ground Ops Costs/flt.	$3,371	$3,964	$3,213
2). Propellant Costs /flt	$35	$54	$638
3). Launch & Mission Ops Costs/flt.	$1,135	$1,776	$574
4). Gnd Transportation & Recovery/flt	$163	$242	$225
5). Fees and Insurance Cost/flt.			
a). Launch Site User Fee/flt.	$86	$63	$13
Public Damage Insurance			
b). RLV Premature Loss Charge/flt	$825	$903	N/A
c). Surcharge for Mission Abort/flt	$505	$590	N/A
Indirect Operations Costs (20%)/FLT	$3,735	$3,889	$3,110
1). Program Admin & Systems Mgmt			
2). Technical Support System			
3). Launch Site & Range Costs			

Figure B.2—Military TAV Cost Model Report, Output

Bibliography

Anseluno, Joseph, "NASA Nears X-33 Pick," *Aviation Week and Space Technology*, June 17, 1996.

Clapp, Capt M.B. (USAF ret.), "Aerial Propellant Transfer TAV," Air Force Phillips Laboratory, briefing to RAND, February 1995.

Daehnick, Maj. C. (USAF), "SPACE LIFT Suborbital, Earth to Orbit, and On Orbit (a SPACECAST 2020 White Paper)," Air University, *Air Power Journal*, Summer 1995.

Elias, A. L., *X-34 Program Overview Presented to the TAV Workshop for Project AIR FORCE RAND*, Orbital Sciences Corporation, April 1995.

Frederic, P. C., and A. R. Sjovold, *Launch Vehicle Cost Model Update*, Tecolote Research Corporation, Cost Driver Identification Task 93-013, prepared for U.S. Army Space and Strategic Defense Command and U.S. Air Force Cost Analysis Agency, 16 August 1993.

Gonzales, D., M. Eisman, C. Shipbaugh, T. Bonds, and A. T. Le, *Proceedings of the RAND Project AIR FORCE Workshop on Transatmospheric Vehicles*, RAND, MR-890-AF, 1997.

Haslett, R., TAV Concepts Briefing, Northrop Grumman, presented to RAND for the TAV Workshop, April 1995.

Hess, R. W., and H. P. Romanoff, *Aircraft Airframe Cost Estimating Relationships: Study Approach and Conclusions*, RAND, R-3255-AF, December 1987.

"Industry Steps Up to Reusable Rocket Effort—OSC, Rockwell Selected to Run X-34 Project," *Space News*, 13–19 March 1995, pp. 4 , 37.

Isakowitz, S. J., *International Reference Guide to Space Launch Systems*, Second Edition, American Institute of Aeronautical Engineers (AIAA), 1991.

Koelle, D. E., TRANSCOST 6.0 *(Statistical-Analytical Model for Cost Estimation and Economical Optimization of Space Transportation Systems)*, (version 6.0), 1995 Edition, TCS-TR-140(95), July 1995.

NASA Marshall Space Flight Center (MSFC) Launch Vehicle Cost Model (LVCM), Planning Research Corporation (PRC), prepared for NASA MSFC, 1983.

Resetar, S. A., J. C. Rogers, and R. W. Hess, *Advanced Airframe Structural Materials—A Primer and Cost Estimating Methodology*, RAND, R-4016-AF, 1991.

Technical Requirements Document for a Military Trans Atmospheric Vehicle (TAV) Advanced Spacelift Technology Program, Phillips Laboratory, Space & Missiles Directorate, February 1995.

Technology Readiness Levels Briefing, NASA Goddard Space Flight Center, 1988.